Bedouin Food Habits:
An Anthropological View

By Dr. Donald E. Crim, PHD
Emeritus, Department of Anthropology
Colorado State University
Ft. Collins, Colorado, U.S.A.

And

Curtis R. Crim, BA

All Rights Reserved
Copyright © 2012 by Curtis R. Crim BA
No part of this book may be reproduced or
transmitted in any form or by any means,
electronic or mechanical, including photocopying,
recording, or by any information
storage and retrieval system without permission
in writing from the author.

ISBN: 978-0-9833732-9-2

Printed in the United States of America

First Printing

This book is dedicated to Anna.

CONTENTS

0. Preface..6

I. Introduction

    Anthropology and the Study of Food Ways....7

    Major Studies of Food Ways....................15

    The Study of Bedouin Food Ways..............20

II. The Food Ways of the Traditional Bedouin

    Introduction: The Bedouin and the Camel....22

    Literature and Sources...........................26

    Foodstuffs and Comestibles.....................29

    Food Preparation and Culinary Techniques...34

    Meals and Feasts..................................39

III. Conclusion

    Incompleteness of Data..........................44

    A Comparative Note..............................47

    Bedouin Food Ways: Chances for Survival..49

IV. Epilogue: Modern Food Ways..................53

References Cited..58

O. Preface
By Curtis R. Crim

My father Dr. Donald E. Crim completed this work, and then passed away in 2007 without having published it.

It was discovered five years after his death by his wife (my mother, Patricia R. Crim).

Because he is not available for consultation, I am publishing this book in as close to an exact reproduction as I can.

Other than the inclusion of this preface and an epilogue which I added, what you read in this book is exactly what he had written, word for word, including, as much as possible, formatting. Obviously, I did correct grammatical errors and misspellings when necessary.

I. Introduction

<u>Anthropology and the Study of Food Ways[1]</u>

Ever since the first ethnographic works of the 19th century, anthropologists have regarded the study of food ways as one of the more common and useful categories of ethnographic observation and documentation. And yet, the reasons for this regard are not entirely self-evident.

Other disciplines and categories of scholarship have, perhaps, a more obvious interest in, and defensible right to this subject matter.

Nutritionists, most obviously, focus their scientific interest entirely within this area. Their interests are somewhat selective, focusing on the biochemistry of food, and its intersection of the psychological needs of the human body for nutritionally adequate sustenance.

---

[1] After much thought, I decided to use "food ways" rather than the more conventional "food habits" to designate the primary topic of this paper. "Food habits" carries the connotation of individual preferences in cuisine, while "food ways" seems to indicate more clearly the collective nature of human food preferences, focused within local cultural traditions.

Gourmets and authors of cookbooks and volumes of related culinary matters are similarly focused within the broad topic of food ways (cf. Brillat-Savarin 1949). In contrast to the nutritionists, these people are primarily concerned with the aesthetics of human gastronomy, with a secondary interest in the proximate technology of sustenance; i.e. that technology which focuses exclusively on the manipulation of foodstuffs from the time they enter the kitchen to the time when they appear on the table. This literature deals primarily with the food ways of the affluent upper classes of the U.S. and Europe, with secondary emphasis on folk cuisines as they are transformed into gastronomic alternatives for the affluent, for whom dining may become a hobby.

Another scientific discipline with an interest in food ways is the field of economics. Or, at least, one might imagine that the distal technology of food ways (that technology which produces and brings the foodstuffs to the kitchen) would be of major interest to economists.[2] In fact, most economists tend to treat foodstuffs as non-distinctive from other commodities, to be produced, transacted, transported, and

---

[2] At least, these should be of some interest to "supply side" economists.

consumed as any other commodity might be.³

Now, then, do the interests of anthropologists in food ways differ from these discussed above? One of the claims usually made for anthropology as a holistic discipline might appear to be partial justification for professional interest in this topic; however, such a claim would equally well justify an interest in any category of human behavior, no matter how obscure or peculiar.

A more compelling reason for this interest is to be found in the uniqueness of the study of food ways as a major intersection between the constants of human biology and the distinctiveness of thousands of local cuisines throughout the world.

It may generally be assumed that the physiology of nutritional needs is uniform among all contemporary human groups, although the presence or absence of the lactose enzyme has been shown to be a major exception to this generalization (Farb and Armelagos 1980:181-184).

---

[3] Malthus was the significant exception. His thesis was taken up, by and large, by demographers rather than economists.

By contrast, local cuisines vary endlessly throughout the world, although these variations tend to follow larger regional and national preferences. Farb and Armelagos present and describe fifteen of these regional patterns, from Japanese to Mexican, which are focused on preferred combinations of seasonings, and thus tend to override local differences in available foodstuffs (Farb and Armelagos 1980:180-181[4]).

Probably even more than sexuality and aggressiveness, the study of food ways dramatically illuminates the interface between biological necessity and cultural inventiveness. Sociobiologists would do well to take note.

We turn briefly to the various topics which have involved anthropologists in the comparative study of food ways.

One of the most obvious of these is the relationship between food ways and sex roles. Men and women, in most human societies, play largely non-overlapping roles in relation to food procurement, preparations, and consumption. For foraging societies, the traditional postulated division of labor

---

[4] Farb and Armelagos' Consuming Passions (1980) is an outstanding recent volume which surveys the broad spectrum of anthropological interests and findings regarding food ways.

between "man, the hunter" and "woman, the gatherer", has become under substantial scrutiny and reformulation in recent years, especially by Feminist anthropologists (of both sexes). However, the basic facts indicating that men generally obtain large animals for eating, while women obtain smaller game and vegetable foods, seem to have survived the reformulation relatively unscathed.

In addition, in numerous societies, men and women eat at different times, the most common practice being that the men are served first, followed by women and children. And while this common practice might be best explained as a social symbol of male supremacy and control, it might also be true in some instances that since women perform the major tasks of food preparation, they and their small children also have extensive access to food for snacking between formal meal times, which men do not have. Thus this feeding sequence might be a simple recognition of the fact that men have probably gone for longer periods without eating.

Another phase of sex-role differences is revealed by differentially allocated cooking duties. When men cook, they generally prepare meat from large game animals, or from large domesticated animals. These foods are usually less frequently available in the diet, and their consumption often an occasion for feasts. Women, by contrast, generally do the bulk of food preparation, and provide more usual but less prestigious foods.[5]

Another topic which has received systematic attention by anthropologists has been the development and transformation of food-oriented behavior in the life-cycle of the individual. From the documentation of food taboos for prospective parents, through nursing, weaning, the learning of food customs in childhood to the food-centered traditions of puberty, adulthood, and old age, anthropologists have assembled a massive archive of food-related data (Farb and Armelagos 1980:71-91).

A related topic is that of the nexus between food ways and religion. Sacred

---

[5] The native cultures of Polynesia are a significant exception. In most of these, cooking is done in large, rather than small groups, only a single cooked meal is prepared, and men serve as principal cooks, in that they are in charge of preparing the luau pit, heating the stone and gravel, and assembling the food in cooking pits and piles. Some parts of New Guinea show a similar pattern. Another distinctive pattern is found among the Inuit peoples of the northern periphery of the North American continent. Here, the women prepare all the food (mostly animal protein), while the men procure at least 95% of the food supply (cf. Balicki 1974).

works, from the Vedas to the Torah and the Christian Bible, devote much space and attention to defining the food ways most acceptable to the Deity in question.[6] Similarly, the literature of foraging and tribal societies reveals numerous examples of food taboos, feasting and fasting, sacramental foods, and elaborate patterns of food symbolism, from the grizzly bear and the salmon spirits of the northwest coastal region of native North America to the Lamb of God among Christians.

Physical Anthropologists, in addition to their interest in genetic variation in contemporary populations, have pursued their interest in hominid evolution through the study of ancient food ways, both by the reconstruction of the food ways of fossil hominids, and the study of food ways of contemporary non-human primates.

To the prehistoric archaeologist, the study of the transition from foraging to food producing societies (horticulturalists, pastoralists, and agriculturalists) has been

---

[6] The Qur'an is a major exception. The refusal of food, in the form of the periodically required fast of Ramadan, is one of the pillars of Islam.

pursued with vigor and controversy in recent years.[7]

In conclusion, anthropologists have shared with sociologists, and to a certain extent, with psychologists, an interest in the relationships between food ways and social structure. The highly stratified traditional societies of Polynesia marked the status of the paramount chief with a complex series of food taboos and restrictions, including major limitations of his food, serving and cooking vessels, preparations, and company at the table. In contrast, the more egalitarian societies of traditional New Guinea did not provide either food taboos or special gastronomic privileges for the "big man" type of leader. Cross-cultural research generally supports the conclusion that in stratified societies, the different social strata generally have different access to foodstuffs; the most usually cited example being the Hindu caste system. On the other hand, egalitarian societies usually do not make differential allocation of foodstuffs, beyond the customary division of male vs. female, and adult vs. child.

---

[7] The problem of Aztec cannibalism, in particular, has spawned a sizeable polemical literature (cf. Harner 1977, Harris 1977; Arens 1979; etc.).

Major Studies of Food Ways

In this section, I mention briefly a series of anthropological studies of food ways which have had a major impact on anthropological interest in this topic. This listing is hardly definitive, or even authoritative; however, it does represent studies which appear to me to have historical significance in the development of this area of study.

Boas' monumental ethnographic studies of the northwest coast peoples of the North American continent established a number of precedents for the development of cultural anthropology in the United States, not the least of which was a graded series of emphases within the body of cultural materials. Boas' broad, but by no means undiscriminating interests in cultural data ranked food ways among the more important categories for observation. This is documented in two ways. First of all, Boas spent considerable time assembling an impressive quantity of recipes for the preparation of local foods, especially salmon (Boas, 1913-1914). Next, Boas devoted much time and attention to the study of the potlatch, seeing it as the major means by which status was validated, and rivals vanquished. Boas' description of potlatches emphasizes the huge amounts of food prepared for such occasions, and the elaborate protocol involved in the serving of foods to guests, in the precise order of their relative rank. Also, guests were deliberately served far more than they could possibly consume. Not only was food lavished on guests, but edible candlefish oil was poured on the fire in large amounts, enough to occasionally set the plank house on fire (Benedict 1934:178-179).

The principal value of Boas' work here was the documentation of the complex patterns of food distribution (and destruction) in reinforcing the rankings of individuals in a highly stratified society, as well as in establishing stable relations between clans.

Another major ethnographic endeavor was Malinowski's splendid studies of the Trobriand Islands. In Coral Gardens and Their Magic, he described in great detail the operations of the garden magician, who supervises and schedules every phase of the agricultural cycle, and who provides the appropriate magical spells for each new phase (Malinowski, 1935). Given that the yam crop is at the center of Trobriand subsistence, the ritual elaboration of the magical support system appears to underline the value of this crop. At another point, Malinowski provides an account of the complex pattern of redistribution of the yam crop following harvest, in which the bulk of the crop is contributed to the store-house of the chief, who uses this surplus to compensate followers for services rendered, and to provide generous hospitality for visitors to the village (Malinowski, 1935).

A lesser-known, but equally interesting study of food ways in cultural context is provided by Holmberg's study of the Siriono of Bolivia. The tropical rain-forest denizens eke out a precarious existence on wild game and vegetable foods, supplemented by a small amount of horticulture. According to Holmberg, the Siriono experienced a chronic shortage of the more desirable foods, especially meat and animal fat. The Siriono, unlike so many hunting peoples, lacked traditional patterns for distribution of a game by the hunters, and in consequence, quarreled more or less continually about each other's lack of generosity. Even their dreams reflected this preoccupation; Holmberg noted that dreams of gorging on favorite foods were very common, while erotic dreams were conspicuous by their absence (Holmberg 1950:241-242).

Of the various British social anthropologists that came to prominence in the 1940s, Audrey R. Richards showed the most sustained interest in food ways. Her major study of the Bemba of Rhodesia focused on various aspects of the nutritional process, following the subsistence pattern from the initial clearing of rain forest to the complex problem of allocation of seasonally scarce millet reserves during the hungry season just before the harvest of the new crop. Another interesting feature was her description of the brewing and serving of native beer, occasions which were as important socially as they were nutritionally (Richards 1939).

The period of World War II in the United States saw a new and significant involvement of anthropologists with the practical study of food ways. As Rhoda Metraux narrates it, the Committee of Food, a subcommittee of the Food and Agricultural Organization of the U.S. government, carried out a series of studies of ethnic populations in the United States, in order to make policy recommendations concerning the allocation of foodstuffs made scarce by the wartime conditions. This committee was directed by Margaret Mead, and employed a number of anthropologists, including Metraux and John Bennett.[8] In this case, we

---

[8] Bennett was working on his doctoral dissertation at the University of Chicago prior to this assignment. His dissertation topic was "Subsistence Economy and Food Ways in a Rural Community: A Study of Socio-Economic and Cultural Change" (Bennett 1946).

have a rare instance of a successful committee operation, which produced sound recommendations which were actually implemented by a federal government (National Research Council 1943, 1945; Odyssey Series 1981).

The 1940s marked the high point of interest of U.S. anthropologists in studying food ways. Only one study of note has come out of the post-war period. This is Rappaport's ecological study of the Tsembaga Maring of New Guinea (Rappaport 1968). Rappaport's interest was not in doing a traditional study of food ways, but rather in describing and documenting the relationships between the long-range variations in sizes of pig herds and the seven to eleven year cycle of ritual warfare. In the process of pursuing this topic, Rappaport presents a large volume of information on food production and allocation. In contrast with previous studies, this one raises and endeavors to answer questions of nutritional adequacy of food supplies for different segments of the population.

The Study of Bedouin Food Ways

When my colleague, Jeffrey Eighmy, made the initial arrangements for a number of us in the Anthropology Department at Colorado State University to be allowed to participate in the Saudi Arabian studies being directed by the Consortium for International Department, I conceived of the idea of an investigation of Bedouin food ways. My interest in this subject originated while I was working on a Masters degree in anthropology under the supervision of John Bennett at the Ohio State University during the late 1950s. It was at that time that I first came across the major ethnographic descriptions of the Rwala Bedouin by Musil (1928). It was under Bennett's guidance that I first began to read systematically in the cross-cultural literature on food ways.

Having had a prior familiarity with much of this literature, I approached the study of Bedouin food ways with high, and in some respects, naïve expectations. I had been impressed with the thoroughness of Musil's descriptions, and was hopeful of finding a reasonably rich literature which would extend his descriptions to other tribes and regions, and even add descriptive dimensions which Musil treated only lightly. In addition, I had hoped that the literature in the field of nutrition would produce a series of studies oriented to the biochemistry of traditional Bedouin food ways. As we shall see in the next section, my expectations were unrealistically high, in that the available literature turned out to be much like the Bedouin diet, capable of sustaining life and vigor, but hardly rich or abundant.

II. THE FOOD WAYS OF THE TRADITIONAL BEDOUIN

Introduction: The Bedouin and the Camel

In the minds of many Americans and Europeans, the camels and Arabs are inextricably linked, and frequently form the basis of a demeaning stereotype. Such a stereotype is as inaccurate as it is unfair, since a vast majority of Arab peoples have been urbanites who have had no particular familiarity with camels.

More accurate is the image of the Bedouin as the classic camel nomad. In point of fact, it is undoubtedly true that the vast desert interior of the Arabian Peninsula was unoccupied until the arrival of the ancestral Bedouins, who were already equipped with domesticated camels (Bulliet 1945:45ff).

The camel has a long and interesting evolutionary history prior to the time when it became the principal vehicle of the Bedouin tribes. Its ultimate origin has been traced to the cameloids of South America, where its nearest relatives, the alpaca, the vicuna, and the llama, are still to be found.

During one of the Pleistocene glacial advances, the ancestors of the Asian camel (still the size of large jack-rabbits) drifted across the Bering Strait land bridge from North America to Asia, and thence, spread eastward and southward to the Middle East and North Africa (Bulliet 1975:28-29).

Some time after 5000 B.C., some of the foraging and fishing folk living in the Hadhramaut region of the Arabian Peninsula, who already included wild camel among the game they hunted, apparently initiated the steps that ultimately led to the domesticated camel. Also, it seems that the desire for camel milk was a primary motivator for this course of events. Although precise archaeological evidence is lacking, Bulliet assembles an impressive set of inferential evidence to support this reconstruction. The ancient and well-known metaphor for the camel as the "ship of the desert", found both in the Qur'an and in Bedouin poetry, is inexplicable unless it arose among people who were familiar with both the sea and the desert. Also, the association of camels with jinn in Bedouin folklore shows maritime connections. Camels are said to be one of the four components of jinn along with man, the sea, and the four winds (Bulliet 1975:47-48).

Bulliet's masterful account of the origin and history of the camel is focused on the problem of explaining why wheeled transportation virtually disappeared from the Middle Eastern countries for many centuries following the Roman period, only to be reintroduced during the past few centuries. Bulliet's major thesis is that during this period, the camel was such an efficient mode of transportation throughout the entire region that wheeled transport could not compete effectively. This excellence of camel transport is in turn linked to the development of efficient saddles, both for riding and packing. Bulliet delves in great detail into the development of different regional saddles: the South Arabian saddle, the North Arabian saddle, and the North African saddle, each specialized in certain ways to adapt efficiently to local environmental conditions and specific cultural utilizations (Bulliet 1975, esp. Ch. 4 and 5).

This background information is included in order to provide a correct perspective on Bedouin camel usages, and to remind the reader that although the following discussion is narrowly focused on the camel as a source of food, the wider picture reveals that the camel has played a major role not only in the history of the Bedouin peoples, but also in the history of the entire region of the Maghreb, and that camel transport did much to facilitate the glorious flowering of the classical Arab civilizations, at periods when Europeans were mired in the cultural backwardness of their Dark Ages.

Literature and Sources

"Not all…temperate zones are equally fertile nor do all their inhabitants enjoy a high standard of living. In some regions, the excellence of soil, the good quality of the plants, and the abundance of population assure their inhabitants an abundance of cereals, rich foods, wheat, and fruit. In other regions the soil is so hot that no plants or grasses grow, hence their inhabitants lead a very hard life…

The nomadic Bedouins also come in the same category, for although they get some cereals and rich foods from the plateau, they do so only occasionally and in spite of the opposition of the sedentary dwellers; hence they cannot get enough of these foods to maintain life, much less to enjoy luxury, and must rely on milk, which compensates them for the absence of wheat. Now it is to be noticed that these inhabitants of the desert, in spite of their lack of cereals and rich foods, are sounder in their minds and bodies than the sedentary peoples who enjoy a softer life. Their skins are clearer, their bodies purer, their figures more harmonious and beautiful, their characters more moderate and their minds sharper in understanding and readier to acquire new knowledge than those of sedentary peoples."

From the Muqaddimah of Ibn Khaldun (Issawi, tr. And ed., 1950)

Thus did one of the first-magnitude luminaries of Arab intellectual history regard the Bedouins during the 14th century. It is interesting to note that Khaldun anticipated a major body of theory in European and American social science, namely the Cultural Materialists paradigm, which sees subsistence technology as casual prior to the development of particular forms of social organization and ideologies. One anthropological observer credits Ibn Khaldun with the writing of the first legitimate science of history, centuries before this genre of writing was to appear in European scholarship (Voget 1975:36-38).

We now turn to literature on the Bedouins which can, in some sense, be considered "ethnographic". The basic and most respectable account is Musil (1928) on the Rwala Bedouins, which uses a standard ethnographic format, and covers material in all of the above categories. Another ethnographical account is Cole's 1975 report on the Al Murrah Bedouins of the Empty Quarter, especially useful for its detailed treatment of the yearly cycle of movement from pasture to pasture, and its discussion of the sexual division of labor in food preparation and serving. A further ethnographic account is Katakura (1977), which describes a recent Bedouin settlement in Western Saudi Arabia, and adds a useful historical dimension to the previous treatments. Of the various travelers who left descriptions of their wanderings, that of Dickson (1949) contains the best information on food-habits, and is further valuable for its detailed illustrations of cooking and food storage implements and containers.

Other reports are of lesser usefulness. Burckhardt, writing during the first half of the 19th century, provides helpful information concerning the food ways of the Syrian Bedouins, as well as briefer notes on the use of locusts by Arabian Bedouins, and on circumcision feasts (Burckhardt 1931). Philby (1952) notes occasions when he was offered hospitality by various village and nomad hosts, while Raswan (1947) includes a horrendous account of a journey with the Bedouins during a period of meager food supplies, on which everyone but important guests had to manage in the tiniest of daily rations.

Foodstuffs and Comestibles

For the desert-dwelling, camel-herding Bedouins, the yearly subsistence cycle is divided into long periods of dryness, separated by shorter, less predictable periods of rainfall. The Rwala, aware of the greater abundance and regularity of rainfall in the settled agricultural regions, explained this contrast as follows:

"The angel Gibrin, who rules over the rain clouds, likes neither the Rwala region nor the desert; therefore, he spreads his wings over them, so that it rains only in narrow strips… On the contrary, when he flies over the territory of the fellanin, he folds his wings as close together as possible, and then the rain falls everywhere… The teachers of the Koran … say that he is angry with the Bedouins because they disregard the instructions which he imparted to the Prophet (Mohammed)" (Musil 1928:14).

For the Rwala, the term rabi does not refer to a given season of the year, but only to the occasional abundance of local rainfall (Musil 1928:14). The Al Murrah, by contrast, recognize four or five distinct seasons. The fall, al-asferi, lasts from mid-September to about the end of December. Winter, ash-shita, follows, and is sometimes followed by ar-rabi'ah (probably cognate with Musil's rabi), a period of abundant rainfall. This is followed by <u>as-seit</u>, running from late March to early June. Finally, the period from June to September, <u>al-gaidh</u>, is the period for halting at summer wells (Cole 1975:39).

Of the traditional foodstuffs, camel milk and milk products are of the greatest nutritional importance. Indeed, the primary determinant of the seasonal movements of the Bedouin tribes is the continual search for the best pastures for their milch animals, as is true for most transhumant pastoralists. Milch camels are kept in the immediate vicinity of the camps, and are milked daily. Milk is either consumed fresh, or boiled and soured in a leather sack. Soured, it is either drunk, or dried and made into cakes of cheese. Milk is consumed daily by virtually all members of the camp.

"Milk is the chief nutrient of the Rwala. Many families live on it exclusively for months at a time; they suffer hunger when there is no abundant pasture, and the animals accordingly have little to eat" (Musil 1928:90).

Among sheep-owning Bedouins, the sheep milk can be used to make butter, in addition to being consumed in sweet or sour form. The buttermilk that remains after churning also provides a useful potable beverage (Dickson 1949:402).

Probably the most commonly eaten fruit among the Bedouin is dates, valued for their portability, flavor, and the ability to remain edible for long periods. Among the Rwala, dates occupy a relatively minor place in the diet (Musil 1928:94); however, Dickson reports that in north-eastern Arabia, even the poorest Bedouin eats dates with his camel milk daily (Dickson 1949:89). Dates may be eaten as is, made into a paste with flour, or boiled in butter (Musil 1928:94).

The third major component of the daily diet is cereal grain. Wheat and rice are the most commonly used, and their use depends on the supplies available from farmers with whom the Bedouins trade. Among poorer Rwala, white sorghum is sometimes substituted for wheat. In leaner times, and in drier desert regions, these domesticated grains are supplemented or replaced by a variety of wild grass seed from the <u>semb</u> plant, which are gathered and prepared in the same manner as wheat and sorghum.

Of the wild foods, the eating of locust has attracted the most attention of outside observers. Dickson, speaking of the Bedouin of north-eastern Arabia, states that the varieties of locust most commonly eaten are the carmine or red-colored types, and the Common Yellow Desert variety, with the brown markings (Dickson 1949:448). On the other hand, among the Rwala, the large green varieties are most commonly eaten (Musil 1928:93). Further, Dickson reports that only the females are eaten (Dickson 1949:448), while Musil mentions no gastronomic discrimination based on sex.

In addition to locusts and semb seeds a variety of desert plants are eaten, especially during lean times, or when returning from an unsuccessful raid. The preferred foods for the Rwala are the fruit of the thorny msa shrub, the sap of the rimt shrub, and the ripe fruit of the butum tree. Several additional varieties are gathered for their stalks and tubers (Musil 1928:95). By contrast, mushrooms and truffles are gathered and eaten only during periods of abundant rainfall (Musil 1928:15). Syrian Bedouins gather and consume three varieties of truffle. In a season following abundant winter rains, children and servants gather truffles in great quantities. Each family will gather four or five camel-loads, and while the supply lasts, truffles are eaten more or less exclusively (Burckhardt 1831:60-61).

The hunting of game apparently varies widely among the various Bedouin peoples, although at no time does it appear to contribute significantly to the daily diet. Musil lists ibex, antelope, gazelle, hare, porcupine, wild pig, hedgehog, lizard, and snake among the game animals of the Rwala.

Gazelle hunting and falconry appear to be recreational activities of wealthy and prominent <u>shaikhs</u>, and appear to have no more dietary significance than does fox-hunting among upper-class Britons.

Meat from domesticated animals is a less frequent item in the diet, and is largely confined to banquets given by the wealthier <u>shaikhs</u> in honor of distinguished visitors, and ceremonial occasions such as circumcisions, and funerals. Camel is most relished; however, many Bedouin groups raise sheep and goats, both for local consumption and for selling to neighboring peoples.

In addition to water and milk, coffee is the most important beverage. Indeed, the preparation of coffee marks the beginning of each day, while the ceremonial brewing of coffee is a major part of the entertainment and welcoming of male guests.

Food Preparation and Culinary Techniques

Unsurprisingly, the major division in the allocation of the tasks of food preparation is based on the sexual division of labor. A statement from Cole concerning the Al Murrah, based on field work done in the 1960s, appears to reflect accurately this traditional allocation:

> "There is a clear sexual division of labor, but both sexes perform activities mainly associated with the other sex under certain conditions. For example, women cook all meals consumed at home in their tent, although men traveling away from the herds cook their own food even when accompanied by women. ...Men are mainly concerned with herding, milking, and watering camels, but women will perform any of these tasks. ... Both men and women attend markets and buy and sell products. Men alone hunt and slaughter and skin all animals for feasting, and serve meals to their guests" (Cole 1975:79).

Preparation of milk products begins with milking, a daily activity. The Bedouins have worked out a series of precise strategies for keeping the female camel fresh for as long as possible, and for keeping the young from nursing at inappropriate times. Not only are the young brought to the females for limited times each day, but also, the female's teats are secured to a stick to prevent unauthorized nursing. The fresh milk is either consumed directly, or soured for storage and further processing. The sour milk (leben) can be kept for several days for drinking, or further processing by boiling, after which the curds are extracted and pressing into small cakes, which can be saved for extended periods. Sheep's milk is treated in much the same manner; however, in addition to the above treatment, it can be churned into butter and buttermilk, using as churn the leather bag in which it was soured (Dickson 1949:402). Camel's milk apparently contains too little butterfat to be used in this way, or else, like goat's milk, it may contain fat so finely dispersed that it does not yield to hand churning methods.

Among the cereal grains, rice is apparently prepared by the familiar boiling process, although no account of this is contained in the sources at hand. Wheat, sorghum, and wild grass seeds are subjected to more elaborate processing. Grain is ground daily, either in a wooden mortar, or in a hand-turned quern, made by the women from pairs of lava or basalt stones. The crushed and pulverized grain is most commonly prepared as gruel by being boiled to a thick paste with water and salt. On rarer occasions, the flour is made into a dough with water and salt, and small portions are formed into large, thin sheets (somewhat like pizza dough), then baked quickly on a convex sheet of iron which is heated over an open fire. This method may be varied by adding yeast to the dough and letting it rise over night. The Rwala prefer unleavened bread and prepare raised bread only during the cold season (Musil 1928:90-92). The Syrian Bedouins make an unleavened paste of flour and water, which is baked in the ashes of a camel dung fire. They also make a paste of flour and soured milk, which is boiled in balls (Burckhardt 1831:57-58).

Among the Rwala, camel is the meat most commonly consumed, although it is hardly to be considered daily fare. The best meat is obtained from a female from four to ten years of age. Apparently males are not eaten, or as least, not relished. The camel is slaughtered by having its throat slit. All portions of meat, fat, viscera, and blood are consumed; only the contents of the stomach and intestines are discarded. The meat surrounding the hump is considered tastiest.

The meat is either boiled in an open kettle, or baked, either on a convex iron sheet, a stone slab, or over red hot embers. The fat may either be roasted in portions on a stick, or cut up and boiled in a kettle into suet. The suet is a preferred garnish for other foods, and can be kept in a leather bag for as long as three years (Musil 1928:96-07).

Among the Syrian Bedouins, a lamb is sometimes baked in a pit, in which a fire has been burned to heat rocks which line the pit. They also occasionally boil herbs in butter, which is stored in skins and used as a seasoning for other foods (Burckhardt 1831:240).

They prepare locusts by throwing them live into a kettle of boiling salted water. After a few minutes, they are taken out and dried in the sun. Heads, feet, and wings are removed, and the bodies, after being rinsed of salt, are stored in sacks for later consumption, or eaten immediately. They are sometimes broiled in butter, or mixed with butter and spread on bread (Burckhardt 1831:92).

Probably the most elaborate preparations are reserved for coffee, especially when served to male guests in the men's side of the family tent. Of all the food utensils, only those used in the preparing and serving of coffee are kept in the men's side of the tent. These consist of a set of three or more coffee pots (for boiling, brewing, and serving), a long-handled roasting pan, a wooden slipper for cooling the coffee beans after roasting, a brass mortar and pestle for grinding the beans, a brass case which holds the handle-less coffee cups, a tripod for brewing the coffee, wooden implements for raking the coffee beans and stirring the brew, and a wad of plant fiber which serves as a filter to hold back the coffee grounds when the beverage is poured (Dickson 1949:84-85). Ordinarily, the host would prepare the coffee himself, as guests and men of the household wait expectantly; however a wealthy <u>shaikh</u> might delegate this task to a slave or a younger man of the household. The coffee itself is usually flavored with cardamom and/or saffron, and in the northeast, following the Turkish custom, sweetened with castor sugar.

Meals and Feasts

Daily meals are apparently not especially standardized among the various Bedouin peoples. Seasonal variations in food supplies have an impact on both quantity and frequency of meals. Dickson generalizes somewhat hastily as he states:

"The well-to-to Badawin…has one square meal a day, usually in the evening. This consists of plain rice cooked in <u>semen</u> (ghi), a few dates dipped in butter, some <u>leben</u> (buttermilk), finished up with coffee. In the early morning, he starts the day with some <u>leben</u> and dates, also coffee. …The poor Badawin… gets one meal a day, consisting of dates and camel's milk, with a very occasional dish of plain rice, or a piece of bread. …The poorest Badawin, such as the herdsmen who tend the sheep or camels of the more fortunate, gets nothing but camel's milk. Very rarely, he gets dates also, and still more rarely, a bit of rice and meat when his <u>shaikh</u> or employer has a guest, his share being the entrails of the sheep that has been killed" (Dickson 1949:189).

This pattern of meals may be contrasted with Musil's statement concerning the Rwala, who regularly eat two meals a day. The main meal occurs after sunset. Shortly before noon there is a meal of leftovers from the previous night's supper. Breakfast is minimal, consisting merely a grain of salt, a morsel of bread, or a drought of milk. On long marches, lunch is usually omitted. "The Rwala know that they will eat only after sunset, and are grateful to Allah if he gives them the chance to eat their fill at least once a week" (Musil 1928:86-87).

Unhappily, the sources at hand provide little information concerning meals for women and children, except for the mention of the fact that at feasts, women and children eat after the men have dined, and separately. However, given that most of the food preparations are done by women, it seems unlikely that they would regularly go short on rations while the men were well fed. There are no indications that the men maintain close supervision over women's activities, as would be necessary if they were seriously interested in limiting the women's food intakes.

Also, despite Dickson's statement that the poorest herders have the most inadequate diet, the fact that the herdsmen have the most continuous access to the herds would seem to guarantee them the most readily available supply of fresh milk. The tying of camel's teats with a stick might deter a hungry baby camel, but hardly a hungry herdsman.

Feasting provides a major contrast to the patterns of daily food intake. Given that hospitality is strictly enjoined on all Bedouin men, rich or poor, the arrival of male guests will inevitably call for the best, most lavishly prepared foods available, and the greatest ceremony in preparation and serving, depending, of course, on the relative status of the guests. Even a poor Bedouin will offer a guest a meal of cooked rice, dates, and sour milk. A wealthier <u>shaikh</u> will predictably offer a large copper platter of boiled rice or wheat gruel or bread, surrounding an entire sheep or lamb, or a large pile of roasted or boiled camel meat. Similarly, on such occasions, the coffee ceremony will be seen in full flower. Under the watchful eyes of guests, the host will carefully roast the beans over the fire, cool them in the wooden slipper, pulverize them with many flourishes in the mortar, brew the coffee, and finally, he will serve each guest and member of the gathering several small portions of coffee in special cups reserved for the occasion.

Special ceremonial manners are also enjoined on the guest. He must not eat until specifically invited by the host, must swallow the balls of dough without chewing, and must carefully eat his way through the cereal food until he reaches the central meat portion, whereupon he must tear the meat from the bone with his right hand alone, and never chew on the bone. If the company is large, the guests may be fed in two, or even three groups. When all have dined, the remainders are given to servants, women, and children (Musil 1928:97-98).

Another category of ceremonial meals marks special religious or life-cycle events. Generally, Bedouins appear to ignore occasions which for Europeans would be marked by special meals. Islamic peoples generally do not utilize Holy days as feast days, as is found in the Judaeo-Christian traditions. The holy period of Ramadan is marked by fasting from sunrise to sunset, and rarely seems to call forth special nighttime foods. According to Cole, the Al Murrah observe Ramadan scrupulously, even when it falls during summer months, causing considerable hardship (Cole 1975:130). Similarly, the only events of the life-cycle that appear to have feasting associated with them are male circumcision and death. Dickson makes brief mention of a feast for mourners shortly after a death, but provides no details. Similarly, he describes the circumcision feasting for a boy between the ages of three and a half and ten years of age. Feasting is to continue for the number of days which corresponds to the boy's age in years. Burckhardt describes a circumcision feast, at which five or six sheep are killed and served to assembled male guests, after which the actual circumcision takes place. The women of the encampment gather outside the tent and sing during the men's banquet (Burckhardt 1831:87-88).

III. Conclusion

Incompleteness of Data

A major limitation of this project has been the incompleteness of the available information. The first aspect of this is that of geographical coverage. There are many Bedouin tribes scattered throughout the Arabian Peninsula, yet only a small handful have had information about food ways published. A further weakness is the spread in time of the available reports. Syrian Bedouins are represented only by Burckhardt's 1831 report, while reliable information on the Rwala is presented only in Musil's 1928 account. Thus, when we encounter differences between these two groups, we are unable to determine whether these are geographical or historical differences. The impressionistic nature of the conclusions is largely a result of the general thinness of the data.

Another aspect of the incompleteness problem is the general lack of attention to women as consumers of food, and indeed, the absence of descriptions of women's activities as preparers of food, or as hostesses. Undoubtedly, this is partially attributed to the fact that most of the accounts used were written by men, as well as the fairly complete separation of men's and women's activities in the daily lives of the Bedouins, a separation which would make it difficult if not impossible for outside male observers to gather information on women's activities. Cole deals frankly with the male bias of his account, and calls for further studies by male-female teams of anthropologists to remedy this imbalance (Cole 1975:80).

Katakura's recent study is the first to be carried out by a trained anthropologist; unfortunately, from the point of view of this study, she had little or no interest in food ways (Katakura 1977).

Finally, there is the lack of studies of Bedouin food ways by nutritional scientists. I was more than surprised that I was unable to find any nutritional studies focused on the Bedouin, or indeed, on any of the nomadic peoples of the Middle East. At one point, I consulted a colleague in the Department of Food and Nutrition at this university. She confirmed my suspicion that there simply was no literature. This is unfortunate, in more than one way. First, it means that we have no way of knowing if there are any health problems among the Bedouins attributable to inadequate diet. Various observers have remarked on the uncertainty of yearly food supplies, and frequent periods of hunger, on one hand, and on the general appearance of good health of the Bedouins, on the other. Second, it means that programs oriented to changing traditional Bedouin food habits are poorly informed as to the nutritional advisability of their goals. One might surmise that nutritionists are more drawn to studies among people who stay in one place, than to people who are continually on the move, where the control of conditions for observations is much more difficult. On the other hand, anthropologists have managed to do significant work under just such conditions, and there is no inherent reason why nutritionists could not do likewise.

A Comparative Note

Originally, I had planned to include a section in which the food ways of the Bedouins would be systematically compared with other peoples whose life-ways resembled those of the Bedouins. I first considered comparing them with peoples whose lives were organized significantly around a single animal species, such as the cattle-herding peoples of East Africa, the Nuer, the Karimojong, and the Watutsi, for instance. Certainly the poetry of the Bedouins, which makes elaborate use of camel imagery, is paralleled by the extensive musical texts of the East African pastoralists which use cattle imagery in the same way, and to the same extent. It is interesting to note that in both areas, East Africa and Arabia, there is a major emphasis on romantic love, and images of the loved one as a cow or camel.

Then, I realized that the Bedouins were culturally involved with both camels and horses, and that a more appropriate comparison might be with peoples who similarly lived with and utilized two species. The most obvious candidates would be the Plains peoples of native North America, who prized their horses as much as the Bedouins prized theirs, and who depended on the vast herds of bison for subsistence even more than the Bedouins depended on camels for food. Now the Plains Indians did not ride bison, so the comparison has its limits. However, there are again some interesting similarities. Warfare and raiding were major cultural foci in both areas, and the role of the warrior was similarly honored. In both regions, social organization was egalitarian and non-stratified, while the technologies were both simple and adapted to mobility. A Plains tepee and a Bedouin tent do not particularly resemble one another, however, they both provide ample protection from the elements, and are easily packed and moved to a new location.

Finally, however, I decided that comparisons of this sort really go beyond the scope of this report, and ought to be saved for a separate paper.

Bedouin Food Ways: Chances for Survival

In an important sense, historical change has already made major inroads into the traditional dietary of the Bedouin peoples, and the details of daily food intake and preparation, as found in the accounts of Burckhardt and Musil, have ceased to exist in their traditional forms. Vast historical changes have swept over Saudi Arabia during the past century, transforming it from one of the most remote, traditional, and mysterious (from the European point of view) countries to one of the most powerful and influential countries in the world at present, controller of vast energy reserves, a stable political presence and power broker in the roiled sea of Middle Eastern politics, a country modernizing on the Western model at a scope of pace literally without historical precedent, blessed with political leadership better educated and more astute than any other nation in that region of the world. Changes of such a scope could hardly fail to make their impact on the nomadic peoples of the Arabian Peninsula, and indeed, they have not.

The Bedouins have been most sharply affected by two major trends, the sedentarization program, initiated by King Abdul Azziz during the 1930s, and the post World War II program of modernization and industrialization. Both have combined to induce the Bedouin to settle in permanent villages, to shift away from full time herding towards more urban types of employment, such as taxi and truck driving and military service, and to gradually abandon their traditional subsistence economy for integration into the urban food ways of the majority of the Arabian population. Recent anthropological studies have tended to focus on economic and ecological changes (Cole 1975; Katakura 1977; and Lancaster 1981), and pay relatively little attention to the proximal food ways detailed in the older accounts. Cole's, however, does provide some tantalizing hints as to the food ways of the Al Murrah during the early 1970's.

Camel milk still appears to be at the center of the Al Murrah diet, in spite of other changes in the periphery. Next to milk, dates still continue to be prominent. Traditional separation of men's and women's dining continues to be observed, as is the ceremonial brewing of coffee in the men's side of the family tent. Hot camel milk spiced with ginger is drunk frequently by men between meals (Cole 1975:136-163).

The situation of the Al Murrah could hardly be taken as typical for a majority of Bedouin peoples. The Al Murrah, after all, are and have been long-distance camel nomads, occupying the Rub' al-Khali, the most arid portion of the Arabian Peninsula, somewhat more protected from pressure to sendentarize than many other Bedouin groups. One would thus expect their lifestyle to be among the most conservative to be extant. On a broader scale, one notes that camels as transportation seem to be diminishing rapidly in importance. Camels as sources of meat still appear to have some significance, even though camel meat appears now to be the food of the urban poor. Bulliet makes a case for the retention of camel meat as a nutritionally and ecologically sound source of protein. He points out that only the camel can make use of the forage provided by the central desert regions, and that they thus do not compete with other edible species (sheep and goats) for limited food supplies (Bulliet 1975:264-268).

Initially, I had planned to conclude this report with a series of recommendations, to be directed at appropriate ministry to the Saudi Arabian government, concerning the best policies to be pursued concerning the food ways of the Arabian Bedouins. On more mature thought, I have decided that this was a misguided notion, for the following reasons:

1. I suspect that interest in Bedouin food ways, per se, is a very low-priority item with the Saudi government. Their major concern is the overall modernization and

industrialization of the country, and the chronic lack of indigenous workers in the labor force, which makes necessary the vast employment of foreigners to accomplish the tremendous work that needs to be done.

2. There may exist some sentiment in favor of retaining at least token populations of Bedouins in their traditional life-style, given the enduring symbolic significance of this life-style to many generations of urban Arabs.

3. It would be presumptuous for me to be offering advice to the Saudi government. After all, I neither speak nor read Arabic, and have never been to the Middle East.

4. There are times when leaving well enough alone turns out to be the best policy (Creative Indifference). After all, despite the hordes of pick-up trucks and luxury vehicles that ply the roads in both city and countryside in Arabia, the historical changes that have led to this situation are not necessarily in the best interests of the Bedouins, or even, of the nation as a whole. When your Toyota pick-up breaks down in the middle of the Rub' al-Khali, you can't eat it.

IV. Epilogue:

Modern Food Ways

By Curtis R. Crim

I was actually disappointed by my Dad's decision to *not* make his recommendations to the Saudi government in his conclusion. By writing this paper, he, in my opinion, clearly demonstrated that he cares about the Bedouin peoples and concern for their food ways and the nutrition that their culture supplies them.

I am not an anthropologist (nor anywhere near as eloquent as my father), but am concerned about the changes to the food ways of not only the Bedouin, but all peoples in the world, especially the native populations.

The "Saudi government", as my father put it, is not the *legitimate* government of Arabia. They were put in power by Shell Oil using the U.S. military in the 1950s. The Saudis are oil-rich billionaires, and like all oil-rich billionaires care nothing for the peoples of their nation other than to enslave them and profit off of their toil, suffering, and death.

The food ways of *all* cultures in the world are now threatened and being destroyed by American food and beverage producing corporations, which are tools used by billionaires to poison the peoples of every country and allow *their* medical industry to profit off of the suffering and death of the poor as well.

The "food ways" of current day American, and indeed, all Western culture are based on mass-manufacturing of food. Mass manufacturing is great when producing tools, such as computers and devices based on digital technology. However, to have a nation's food supply dependant upon computers and by extension electricity is *insanity*.

The current day American food ways are based on computers turning the massive energy resources provided by fossil fuels into foodstuffs for the peoples of many nations. These foodstuffs not only provide little in terms of nutritional sustenance, but are almost exclusively infused with a dangerous toxic narcotic substance known as HFCP (High Fructose Corn Poison) which is an addictive form of heroine refined from mass-produced and genetically engineered Monsanto corn.

I *do* have recommendations for the Saudi government:

1) Make all products containing HFCP illegal in your country, especially soda pop like Coke, Pepsi, and Mountain Dew.

2) Stop consorting with American corporations.

3) Encourage the peoples of Arabia to adhere to traditional methods of producing and manufacturing foodstuffs.

In America, we are now experiencing a diversification in terms of the production of nutritional foodstuffs. Many people appear to be retarded as they continue to purchase food mass-produced by mega-corporations. On the other hand, there is now a growing movement that supports the commitment to the production and consumption of organic and home-grown foods. Many are waking up as to the *low* quality of the products and services offered by billionaire-owned mega-corporations.

I believe that the only foodstuffs that you should trust, consume, and feed to your family are the ones that you can produce yourself. That might seem to be more of a challenge to those living in urban environments, but anyone with an apartment can raise hydroponic vegetables, and anyone with a house can build an add-on greenhouse, even if he hasn't the yard-space to grow a garden.

I recommend that every reader start his own garden, build a greenhouse, and consume only foods grown or raised at home, or by someone else whom he knows personally. I suggest that products and services provided by U.S. mega-corporations should not be purchased or used by anyone. If one is in need of a product or service, he should create or provide it himself. I recommend that one should *never* consume any foodstuffs that contain HFCP.

I also recommend that the reader do some ethnographic research of his own. Spend one half-hour watching the individuals who go in and out of a convenience store (especially in the morning). Collect and record data observed based on the body weight of customers, and the products that they purchase.

My observations indicate a clear correlation between obesity and the consumption of large amounts of foods containing highly refined flour, sugar, and HFCP. These foodstuffs are not only non-nutritious, but are also extremely destructive to the liver, brain, pancreas, spleen, and gall bladder. HFCP is also the primary cause of diabetes and brain damage world wide, and decreases the ability of the human brain to perform higher cognitive functions.

So, the next time you see a retarded, enormous human mountain of redundant protoplasm, note whether it is carrying a Mountain Dew.

REFERENCES CITED

Arens, W.
1979   The Man-Eating Myth: Anthropology and Anthropophagy. Oxford, New York, Toronto, Melbourne: Oxford University Press.

Balicki, Asen
1974   The Eskimo: The Fight For Life. A documentary film of the Netsilik Eskimo. Written and narrated by Asen Balicki.

Benedict, Ruth
1934   Patterns of Culture. Boston: Houghton Mifflin.

Bennett, John W.
1946   Subsistence Economy and Foodways in a Rural Community: A Study of Socio-Economic and Cultural Change. Unpublished doctoral dissertation, University of Chicago.

Boas, Franz
1913-1914   The Ethnology of the Kwakiutl. U.S. Bureau of American Ethnology, Thirty-fifth annual report, Washington D.C.

Brillat-Savarin, Jean Anthelme
1949   The Physiology of Taste, translated by M.F.K. Fisher. New York:Heritage Press.

Bulliet, Richard W.
1975   The Camel and the Wheel. Cambridge: Harvard University Press.

Burckhardt, Richard W.
1831   Notes on the Bedouins and Wahabys Collected During His Travels in the East. Volume I. London: Henry Colburn and Richard Bentley.

Cole, Donald Powell
1975   Nomads of the North: The Al Murrah of the Empty Quarter. Chicago: Aldine.

Dickson, H.R.P.
1949   The Arab of the Desert: A Glimpse into the Badawin Life in Kuwait and Saudi Arabia. London: George Allen & Unwin.

Farb, Peter and George Armelagos
1980   Consuming Passions: The Anthropology of Eating. Boston: Houghton Mifflin.

Harner, Michael
1977   "The Ecological Basis for Aztec Sacrifice,"   AMERICAN ETHNOLOGIST 4 (1):117-135, February 1977.

Harris, Marvin
1977   "The Cannibal Kingdom," pp. 97-119 in Harris, Marvin, Cannibals and Kings. New York: Random House.

Holmberg, Allan R.
1950   Nomads of the Long Bow: The Siriono of Eastern Bolivia. Smithsonian Institution, Institute of Social Anthropology, Publication No. 10, Washington, D.C.

Issawi, Charles, translator and arranger
1950   An Arab Philosophy of History: Selection from the Prolegomena of the Ibn Khaldun of Tunis (1332-1406). London: John Murray.

Katakura, Motoko
1977   Bedouin Village: A Study of a Saudi Arabian People in Transition. Tokyo: University of Tokyo Press.

Lancaster, William
1981   The Rwala Bedouin Today. Cambridge, London, New York, New Rochelle, Melbourne, Sydney: Cambridge University Press.

Malinowski, Bronislaw   1935
Coral Gardens and Their Magic: A Study of the Methods of Tilling the Soil and of Agricultural Rites in the Trobriand Island. London: Allen & Unwin.

Musil, Alois
1928   The Manners and Customs of the Rwala Bedouin. New York: American Geographical Society of New York.

National Research Council
1943   The Problem of Changing Food Habits. Bulletin No. 108, Report of the Committee on Food Habits, 1941-43. Washington D.C., October 1943.

1945   Manual for the Study of Food Habits. Bulletin No. 111, Report of the Committee on Food Habits, 1941-43. Washington D.C.

ODYSSEY Series
1981   Margaret Mead: Taking Note. A 60 minute video film, part of the ODYSSEY Series, produced by Michael Ambrosino.

Philby, H. St. J.B.
1952   Arabian Highlands. Ithaca: Cornell University Press.

Rappaport, Roy A.
1968   Pigs for the Ancestors: Ritual in the Ecology of a New Guinea People. New Haven and London: Yale University Press.

Raswan, Carl
1947   Black Tents of Arabia. New York: Creative Age Press.

Richards, Audrey
1939   Land, Labour and Diet in Northern Rhodesia. London, New York, Toronto: Oxford University Press.

Voget, Fred W.
1975 A History of Ethnology. New York, Chicago, San Francisco, Atlanta, Dallas, Montreal, Toronto, London, Sydney: Holt, Rinehart and Winston.

www.ingramcontent.com/pod-product-compliance
Lightning Source LLC
Chambersburg PA
CBHW060721030426
42337CB00017B/2956